Contents

ALL ABOUT TOILETS 2

BATHROOM BASINS 13

KITCHEN SINKS 18

BATHTUBS AND SHOWERS 24

WATER SUPPLY 28
(HOW TO SOLDER)

DRAINAGE SYSTEMS 35

Printed in Canada.

First published in the United States in 1975 by the Step by Step Guide Book Co., West Valley City, Utah © Step by Step Guide Book Co. 1975, 1981, 1995

IMPORTANT

All of the illustrations in this book show typical plumbing methods - actual installations must be adapted to individual requirements, so follow local plumbing codes in your area.

The authors have made every effort to ensure the accuracy and reliability of the information and instructions in this book. Neither the authors nor the publisher can accept responsibility for misinterpretation of the directions, human error, or typographical mistakes.

HOME PLUMBING

A Step-by-Step Guide

Ray McReynolds

Self-Counsel Press

(a division of)

International Self-Counsel Press Ltd.

IF TOILET IS PLUGGED UP
DO THE FOLLOWING

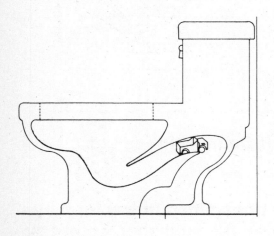

1 Object shown in toilet trapway is usually where toilet plugs up.

By dipping or sponging take the water out of the toilet bowl. Water doesn't need to be turned off.

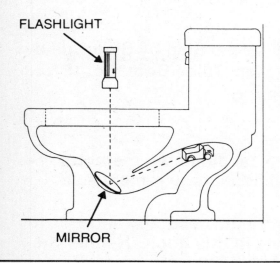

FLASHLIGHT

MIRROR

2 Get a small mirror and flashlight. Put mirror in bottom of toilet bowl. By shining the flashlight into the mirror, it will reflect light up into the toilet trapway and let you see what is obstructing the toilet.

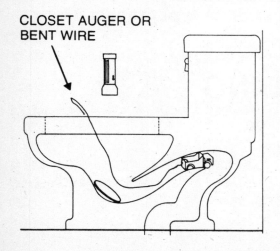

CLOSET AUGER OR BENT WIRE

3 Next get a closet auger or bend a piece of wire and remove object.

How A Toilet Works

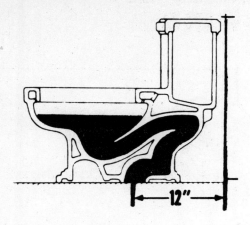

Illustration of a toilet showing the water action in the bowl. The dark area represents the water.

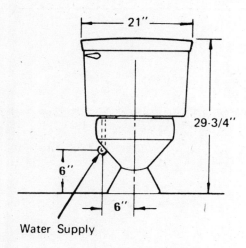

Water Supply

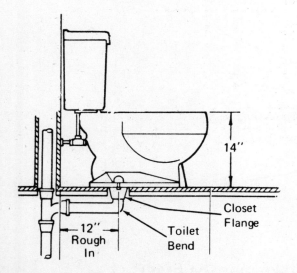

Closet Flange

Toilet Bend

12" Rough In

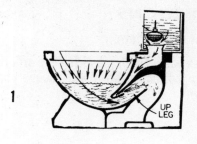

1

UP LEG

2

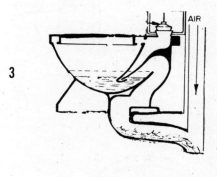

3

AIR

1. Tank ball letting water into bowl, starting the flushing action.

2. Tank ball following the water out of tank.

3. Tank ball closing off water draining into bowl. This completes the flushing action in the toilet.

3

Installing The Toilet Bowl

INSTRUCTIONS

Turn bowl upside down.

Fit the wax gasket with plastic sleeve around opening in bottom of bowl with sleeve portion upright.

Install bowl in proper location with plastic sleeve portion of gasket inside waste pipe and twist lightly to seat gasket.

Tighten flange bolts.

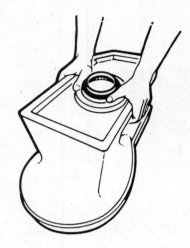

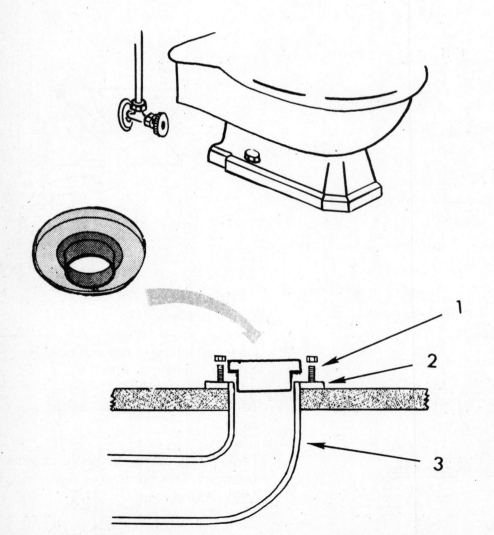

1. Toilet Bolts
2. Toilet Flange
3. Toilet Ell

The screws holding the floor flange down must be brass

How To Install A Toilet On A Cement Floor Using A Toilet Flange

1. HOW PIPE LOOKS NOW—REMOVE BAND AND CAP.

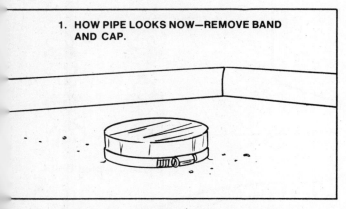

2. STUFF A PIECE OF CLOTH OR PAPER INTO PIPE TO KEEP ANYTHING FROM FALLING INTO PIPE. DRILL SMALL HOLES IN PIPE AS CLOSE TO CEMENT AS POSSIBLE.

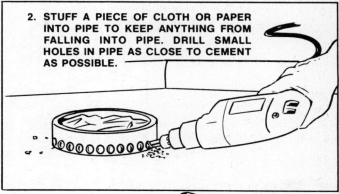

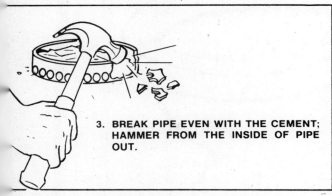

3. BREAK PIPE EVEN WITH THE CEMENT; HAMMER FROM THE INSIDE OF PIPE OUT.

4. CHIP OUT THE CEMENT NEXT TO THE PIPE WITH A CHISEL; CHIP IT ALL AROUND THE PIPE TO THE DEPTH OF THE CLOSET FLANGE.

5. SET THE CLOSET FLANGE ON PIPE AFTER CEMENT IS CHIPPED OUT AROUND PIPE. THE CLOSET FLANGE SITS ON TOP OF CEMENT OR WOOD FLOOR.

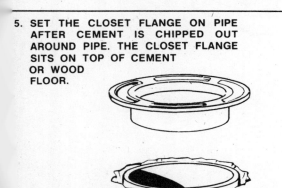

6. TO ATTACH THE CLOSET FLANGE TO THE FLOOR, FIRST STUFF OAKUM WITH A BLUNT SCREW DRIVER OR CHISEL INTO THE SMALL CAVITY BETWEEN THE PIPE AND CLOSET FLANGE; NEXT, DO THE SAME WITH SHREDDED LEAD, ALSO CALLED LEAD WOOL. THIS WILL HOLD THE FLANGE SOLID TO THE FLOOR.

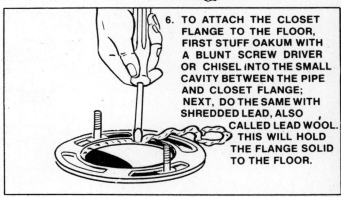

7. INSERT CLOSET BOLTS INTO THE FLOOR FLANGE, AND PACK SOME PUTTY AROUND THEM TO KEEP THEM IN PLACE.

CLOSET BOLT WITH WASHER AND NUT

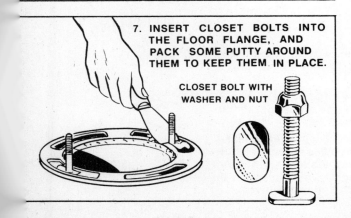

8. A WAX RING WITH PLASTIC COLLAR FITS ON TOP OF CLOSET FLANGE. THE PLASTIC EXTENSION ON THE WAX RING GOES INTO THE DRAIN PIPE, MAKING IT IMPOSSIBLE TO LEAK AROUND THE BASE OF THE TOILET.

Installing the Toilet Tank

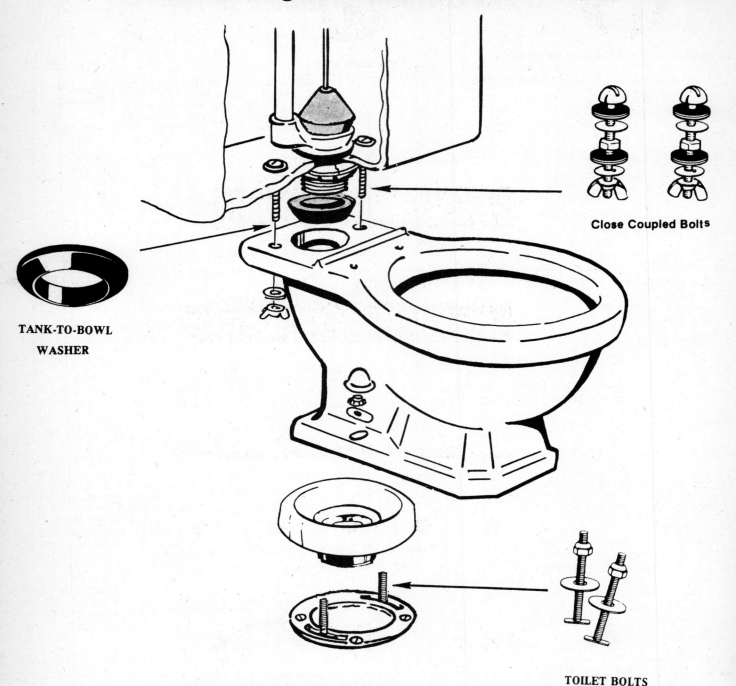

Close Coupled Bolts

TANK-TO-BOWL WASHER

TOILET BOLTS

INSTALLING THE TOILET TANK

Slip the large rubber gasket over the threads on shank of the valve protruding through the bottom of the tank so that it rests snugly against the locknut.

Install channel cushions over the back ridge of the bowl and place the tank on the bowl so that the large rubber gasket fits evenly into the water inlet hole of the bowl.

Place a rubber washer under the head of each bolt and insert the bolts, from inside the tank, through holes in the tank bottom and the bowl's back ledge. Slip washers and hex nuts on the bolts; tighten nuts finger tight. Tighten nuts evenly with adjustable of 9/16 in. open end wrench and screwdriver.

The screws holding the floor flange down must be brass

Toilet Gasket for Close Coupled Connection

Toilet Bowl Wax
For setting toilet bowls.

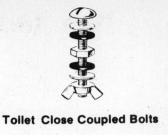
Toilet Close Coupled Bolts

**Toilet Tank
Lower Lift Rods**

REPAIRING TOILETS

**Toilet Tank
Upper Lift Rods**

REPAIRING FAUCETS

Faucet Washer Guide

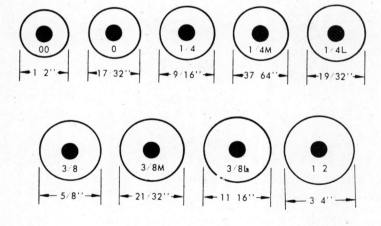

00	0	1/4	1/4M	1/4L
1/2''	17/32''	9/16''	37/64''	19/32''
3/8	3/8M	3/8L	1/2	
5/8''	21/32''	11/16''	3/4''	

ALL DRAWINGS ARE ACTUAL SIZE

BEVELED

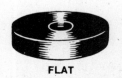
FLAT

How To Hook Up The Water Supply To Toilet

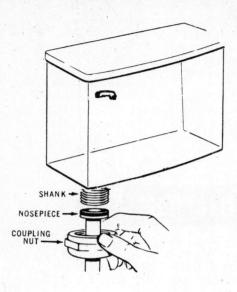

Remove coupling nut from toilet shank and install on flexible supply tube.

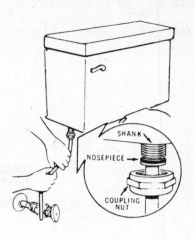

Carefully bend supply tube by hand between the two connections. Make sure not to bend tube within 2" of either end.

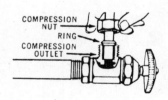

1. Remove compression sleeve and nut from supply stop and install on flexible supply tube.

2. Insert end of flexible supply in outlet end of supply stop and tighten compression sleeve with nut.

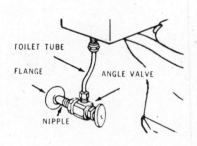

Completed water supply hook-up on toilet.

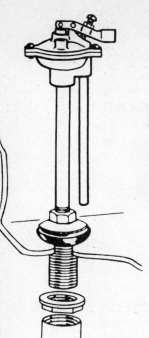

1

Remove coupling nut and locknut from shank.

2

Place shank of ballcock through opening in tank with large rubber cone washer in place, as illustrated.

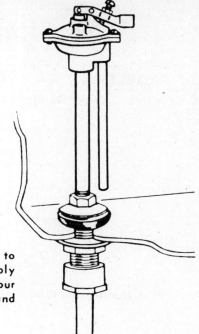

3

Connect water supply to shank. (Should new supply tube be required, see your Dealer for proper size and length)

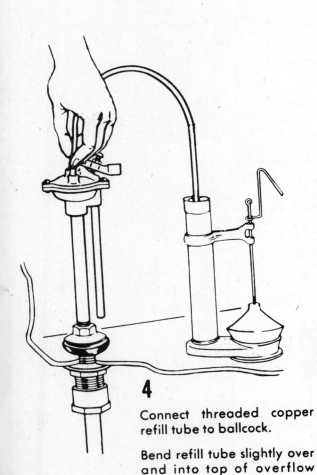

4

Connect threaded copper refill tube to ballcock.

Bend refill tube slightly over and into top of overflow tube.

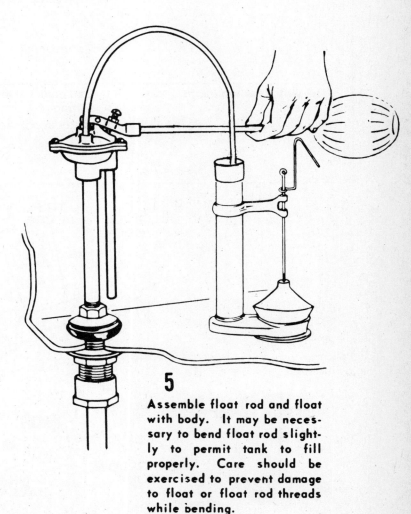

5

Assemble float rod and float with body. It may be necessary to bend float rod slightly to permit tank to fill properly. Care should be exercised to prevent damage to float or float rod threads while bending.

Toilet Troubleshooting Chart

PROBLEMS	CAUSES	REPAIRS
Water in tank runs constantly	1. Float ball or rod catching.	1. Look to see if ball is touching back tank wall. Bend float rod gently to move ball or rod.
	2. Float ball not rising high enough to shut off valve.	2. Gently bend rod down, but only a little.
	3. Tank ball not seating.	3. Check lip of valve, and scrape away any corrosion. Replace tank ball if worn. Adjust lift wires and guide if ball cannot fall straight.
	4. Ball cock valve does not shut off.	4. Washers need replacing, or entire unit needs replacing.
Nonflush or inadequate flush	1. Clogged drain.	1. Remove blockage.
	2. Not enough water in tank.	2. Raise level of water by gently bending float rod up.
	3. Tank ball falls back before enough water leaves tank.	3. Move guide up so that tank ball can rise higher.
	4. Leak where tank joins bowl.	4. Tighten spud nuts or replace spud washer.
	5. Condensation.	5. Install tank cover, drip catcher, or unit to bring hot water to tank.
Tank whines when filling	1. Ball cock valve not operating properly.	1. Replace washers or install new ball cock unit.
	2. Water supply restricted.	2. Be sure cutoff is open all the way. Check for scale or corrosion at entry and on valve.

Ballcock

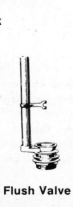

Flush Valve

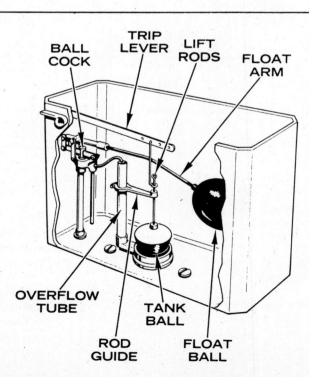

TRIP LEVER

BALL COCK

LIFT RODS

FLOAT ARM

OVERFLOW TUBE

TANK BALL

ROD GUIDE

FLOAT BALL

Flush Valve

Rod Guide

Toilet Rough-in Measurements

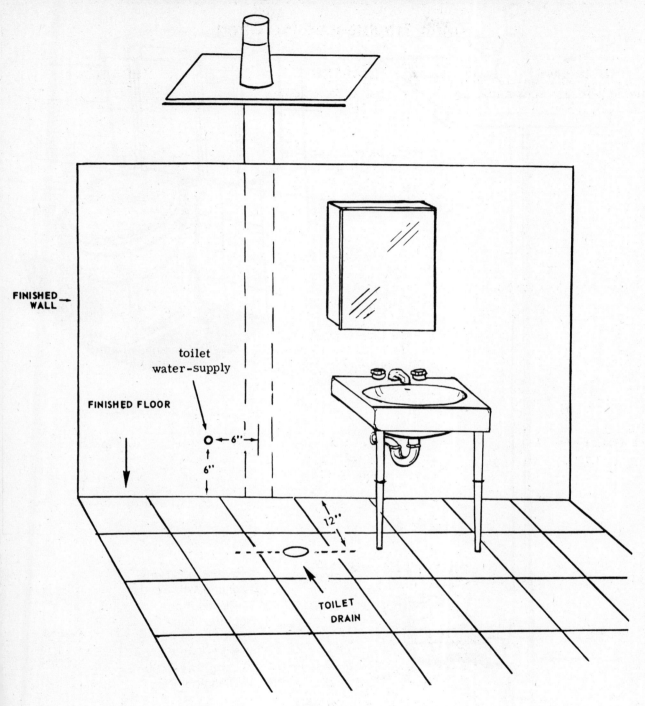

1. Toilet drain is located 12" from center of drain to finished wall.
2. Toilet water supply is located 6" high from finished floor.
3. Toilet water supply is located 6" to the left of the center of the toilet drain.
4. the water closet has to be 15" away from the side wall and 18" away from the front wall. (This may not be a requirement in all plumbing codes).

Toilet Vent And Drain

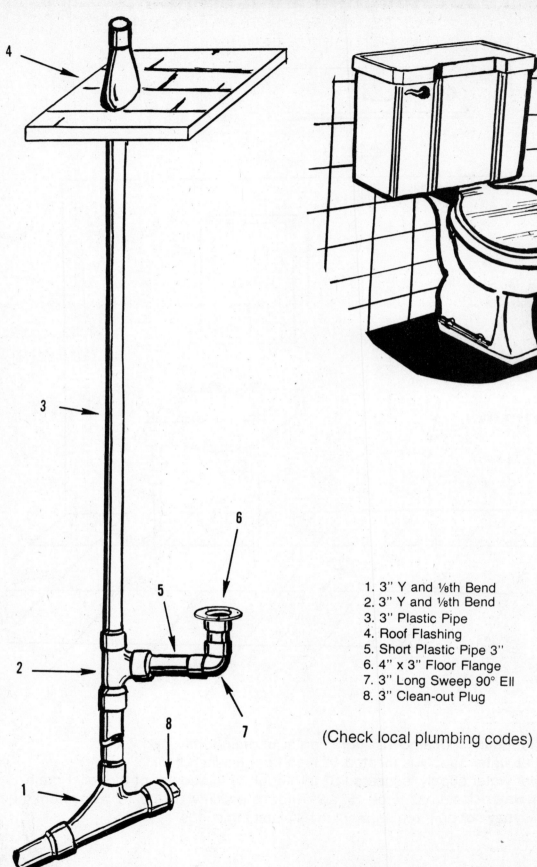

1. 3" Y and ⅛th Bend
2. 3" Y and ⅛th Bend
3. 3" Plastic Pipe
4. Roof Flashing
5. Short Plastic Pipe 3"
6. 4" x 3" Floor Flange
7. 3" Long Sweep 90° Ell
8. 3" Clean-out Plug

(Check local plumbing codes)

All About Faucet Washers And O-Rings

FAUCET WASHER IDENTIFICATION CHART
ALL DRAWINGS ARE ACTUAL SIZE

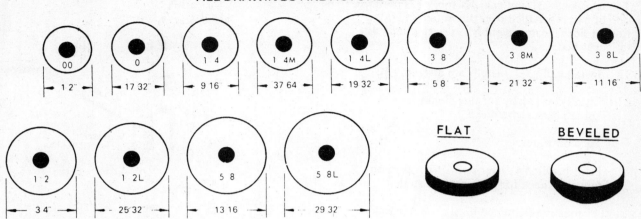

★ All dimensions shown have been accepted by the National Plumber's Association Code.

"O" RINGS

ITEM NUMBER	DESCRIPTION	I. D.	O. D.	THK
R-42	Delta	31/64''	5/8''	1/16''
R-43	Moen, Eljer, Dishmaster, Sayco	1/2''	11/16''	3/32''
R-44	Central Brass	5/16''	7/16''	1/16''
R-45	"Crane"® P.U. Waste, Milwaukee	3/4''	1''	1/8''
R-46	Acme Brass	1/4''	3/8''	1/16''
R-47	Central, Salter, Indiana	7/16''	5/8''	3/32''
R-48	Price-Pfister, Schaible, Mullins, Dishmaster No. 11	9/16''	3/4''	3/32''
R-49	Mullins, Briggs or Republic Schaible, Dishmaster, Central Brass, Kohler	11/16''	7/8''	3/32''
R-50	Sterling, Repcal, Price-Pfister	3/8''	9/16''	3/32''
R-51	Price-Pfister, Arrowhead Brass, American-Standard	5/8''	13/16''	3/32''
R-52	Union Brass	3/4''	15/16''	3/32''
R-53	Moen, Eljer	13/16''	1-1/16''	7/8''
R-54	Mullins, Sterling	7/8''	1-1/16''	3/32''
R-55	"Crane"® stems	27/64''	45/64''	9/64''
R-56	Harcraft, Delta, Repcal, "Crane"®	7/8''	1-1/8''	1/8''
R-57	Universal-Rundle	3/8''	1/2''	1/16''
R-58	Moen	1-1/8''	1-3/8''	1/8''
R-59	Kohler	3/4''	7/8''	1/16''

FULL SCALE DRAWINGS

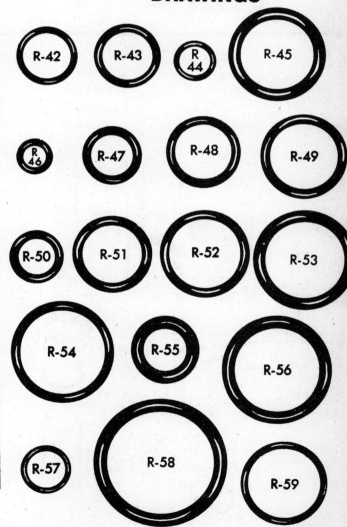

Repairing Ball Faucets

To install ball [3] and cam assembly [2], slot on side of ball is installed over pin in faucet body. Lug on side of cam assembly is inserted into slot in faucet body.

Install ball [3]. Install cam assembly [2].

Install cap [1] by turning clockwise. Using pliers, tighten cap.

Turn on water.

Place handle [4] over stem [6]. Turn on faucet. Check for leaks around stem.

Cap [1] must be tightened until no water leaks around stem [6] when faucet is on and handle [4] is moved.

Using wrench, tighten cap [1] as required. Remove adhesive tape.

Place handle [4] at installed position. Tighten set screw [5].

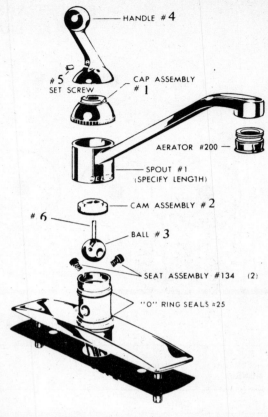

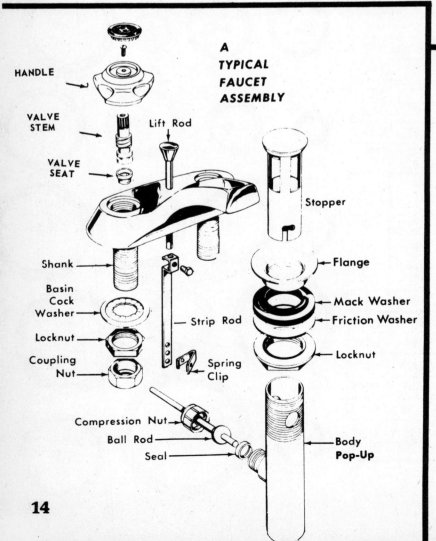

A TYPICAL FAUCET ASSEMBLY

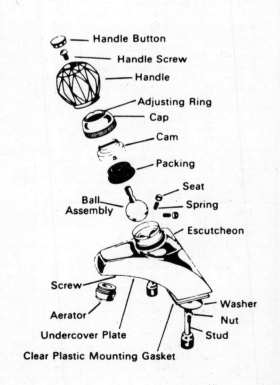

14

Basin Drains And Water Supply

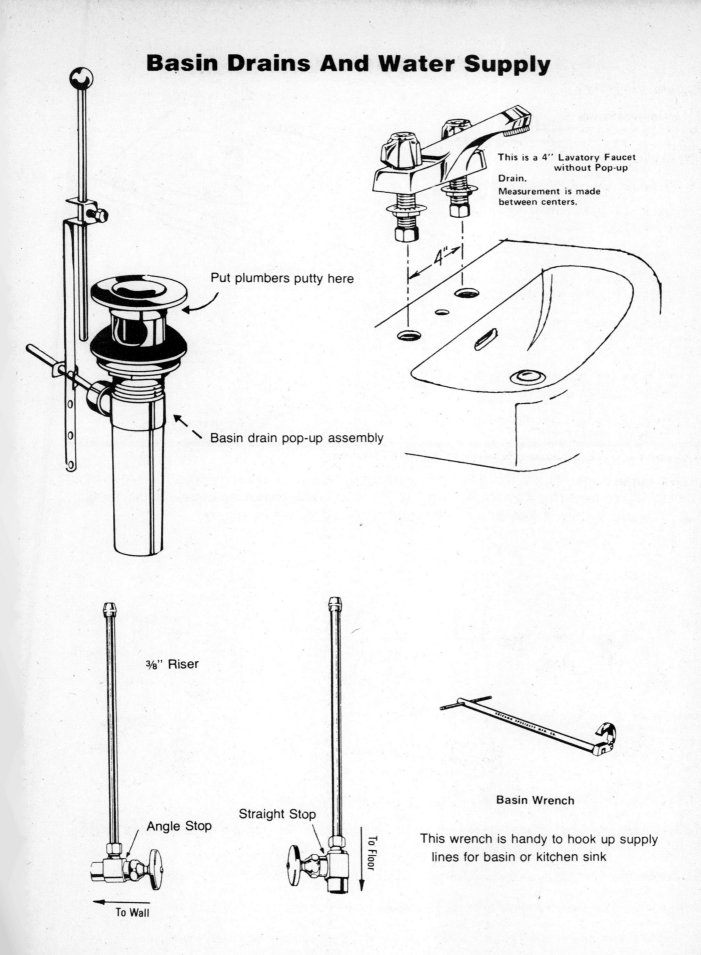

Put plumbers putty here

Basin drain pop-up assembly

This is a 4'' Lavatory Faucet without Pop-up Drain.
Measurement is made between centers.

4"

⅜'' Riser

Angle Stop

To Wall

Straight Stop

To Floor

Basin Wrench

This wrench is handy to hook up supply lines for basin or kitchen sink

Measurements For A Washbasin

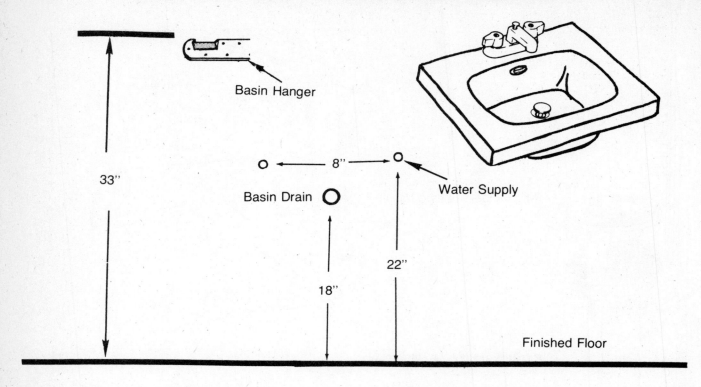

The above measurements are for either a wall-mounted basin or a vanity type. The four most important measurements for wash basins are: 33" from finished floor to top edge of basin hanger; 8" apart on water supply; 18" high for drain; and 18" high for water supply.

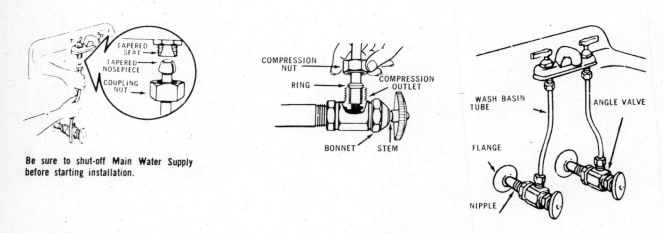

1. Remove tailpiece nut from faucet shank and install on flexible supply tube.

2. Carefully bend supply tube by hand between the two connections. Make sure not to bend tube within 2" of either end.

3. Attach flexible supply to faucet shank and tighten tailpiece nut.

4. Remove compression sleeve and nut from supply stop and install on flexible supply tube.

5. Insert end of flexible supply in outlet end of supply stop and tighten compression sleeve with nut.

How To Connect A Basin Drain To Plastic Pipe

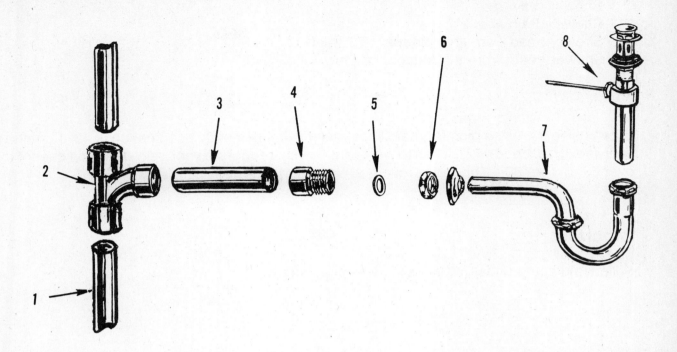

1. Plastic Pipe
2. 1½" Y and ⅛th Bend
3. Plastic Pipe
4. 1½" Hub x 1½" Male Plastic Adaptor
5. 1½" x 1½" Rubber Washer
6. 1½" x 1½" Slip Nut
7. 1½" P Trap
8. 1¼" Pop-up Unit

(Check local codes in your area.)

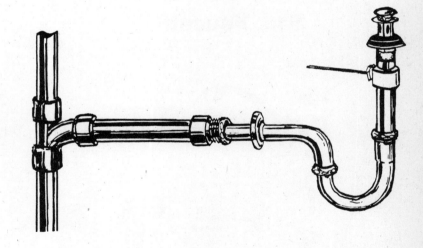

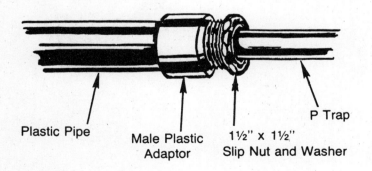

Plastic Pipe

Male Plastic Adaptor

1½" x 1½" Slip Nut and Washer

P Trap

Shower

1. Shower trap size: 2"
2. Shower drain size: 2"
3. Shower drain line: 2".
4. Shower head from finished floor: 78" high.
5. Shower head, from tub faucets to shower head: 48".

Note

When using 1½" size pipe for drain lines, the maximum length it can be run is 42" from the main vent line. If it exceeds 42", it has to be revented with a separate vent. A 2" drain line cannot extend over 60". If over 60," it has to be revented with a separate vent.

Important

Follow plumbing codes when doing any plumbing.

Sink Faucet

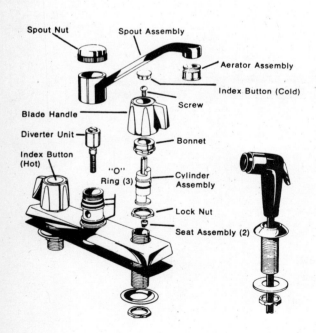

Lavatory Faucet

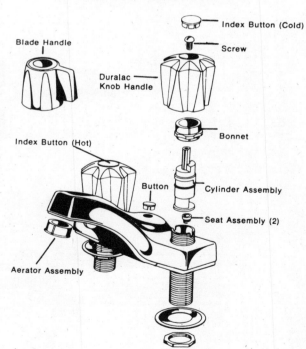

Double Sink Connections

Installation Instructions
End Outlet Tee & Tailpiece

 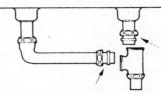 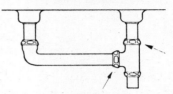

1. Remove nuts and washers from tee.

2. Place nuts and washers on inlet tubes.

3. Slip inlet tubes into tee, tighten nuts to make seals.

Center Outlet Tee & Tailpiece

 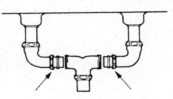 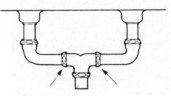

1. Remove nuts and washers from Twin Elbow.

2. Place nuts and washers on inlet waste tubes.

3. Slip inlet tubes into Twin Elbow, tighten nuts to make seals.

Waste Bend Tube with Nut & Washer

 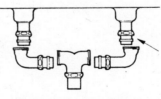 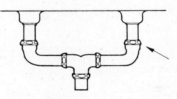

1. Remove nut and washer from waste bend.

2. Place nut and washer on inlet tube.

3. Slip inlet tube into waste bend, tighten nut to make seal.

 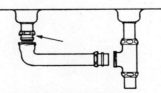 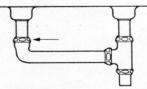

1. Remove nut and washer from waste bend.

2. Place nut and washer on inlet tube.

3. Slip inlet tube into waste bend, tighten nut to make seal.

Kitchen Sink Installation

A. Kitchen sink drain with garbage disposal 16" high from finished floor

B. Kitchen sink without garbage disposal 16" high from finish floor

C. Sink trap size 1½"

D. Sink drain line size 2"

E. Use caulking compound under sink rim.

Kitchen Sink Strainer
Installation Instructions

1. Thoroughly clean around drain opening in sink.

2. Apply ⅛" bead of plumber's putty around drain opening approximately ½" from edge of opening.

3.
Remove nuts and washers and place sink strainer body through opening.

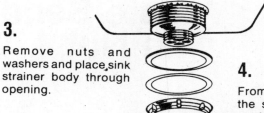

4.
From the underside of the sink place rubber washer and thin metal washer over large threads. Then screw large metal nut tightly against the washers.

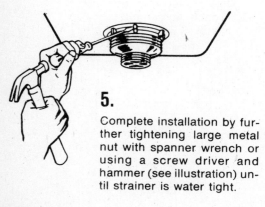

5.
Complete installation by further tightening large metal nut with spanner wrench or using a screw driver and hammer (see illustration) until strainer is water tight.

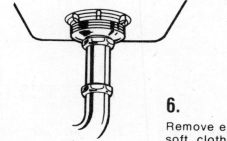

6.
Remove excess putty with soft cloth. Installation is complete.

Drains and Traps for Washbasins, Kitchen Sinks, and Bathtubs

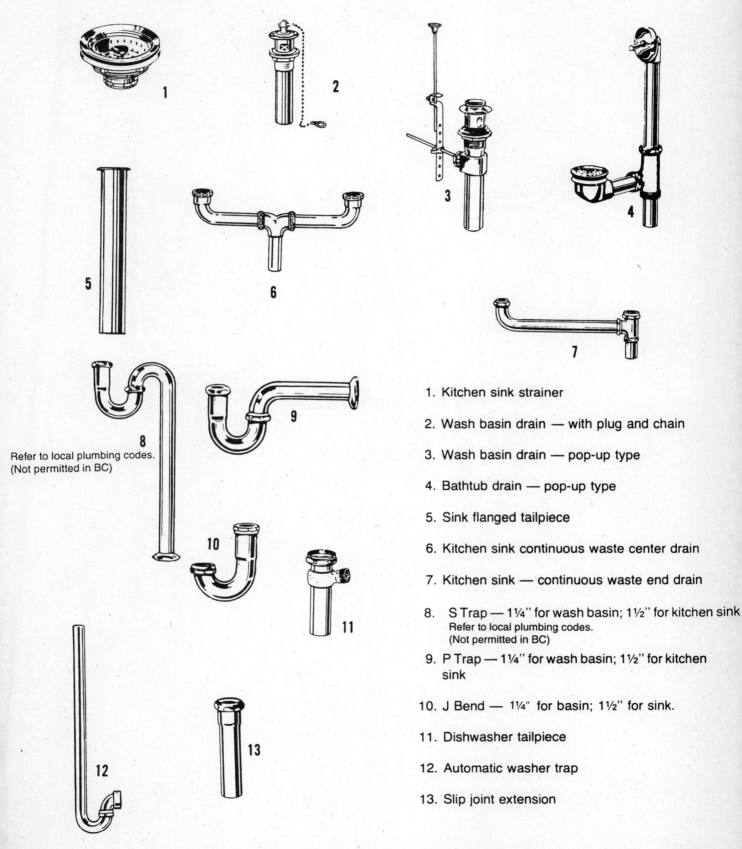

Refer to local plumbing codes.
(Not permitted in BC)

1. Kitchen sink strainer

2. Wash basin drain — with plug and chain

3. Wash basin drain — pop-up type

4. Bathtub drain — pop-up type

5. Sink flanged tailpiece

6. Kitchen sink continuous waste center drain

7. Kitchen sink — continuous waste end drain

8. S Trap — 1¼" for wash basin; 1½" for kitchen sink
   Refer to local plumbing codes.
   (Not permitted in BC)

9. P Trap — 1¼" for wash basin; 1½" for kitchen sink

10. J Bend — 1¼" for basin; 1½" for sink.

11. Dishwasher tailpiece

12. Automatic washer trap

13. Slip joint extension

How To Connect Kitchen Sink Trap To Plastic Pipe

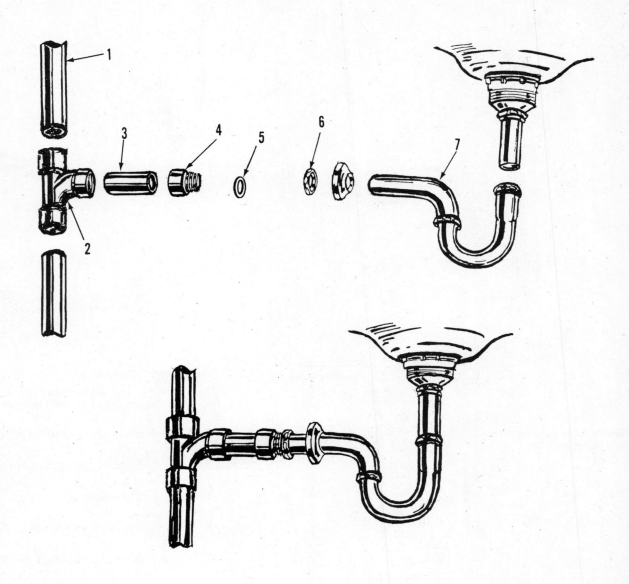

1. 1½" Plastic Pipe
2. 1½" Y and ⅛th Bend
3. Short piece of plastic pipe
4. 1½" Hub x 1½" Male Adaptor
5. 1½" Slip Joint Washer
6. 1½" Slip Nut Washer
7. 1½" P Trap

Connecting Water Supply And Dishwasher To Kitchen Sink

Water To Kitchen Sink

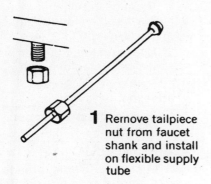

1 Remove tailpiece nut from faucet shank and install on flexible supply tube

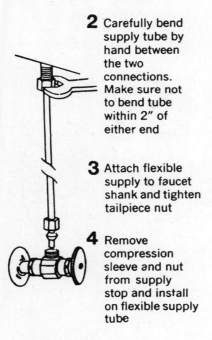

2 Carefully bend supply tube by hand between the two connections. Make sure not to bend tube within 2" of either end

3 Attach flexible supply to faucet shank and tighten tailpiece nut

4 Remove compression sleeve and nut from supply stop and install on flexible supply tube

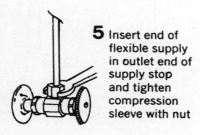

5 Insert end of flexible supply in outlet end of supply stop and tighten compression sleeve with nut

Dishwaster Connection To Kitchen Sink

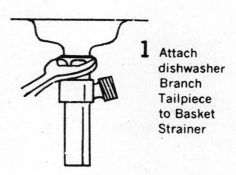

1 Attach dishwasher Branch Tailpiece to Basket Strainer

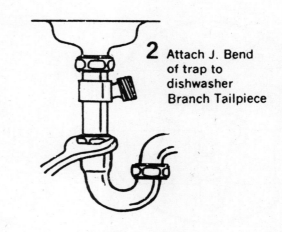

2 Attach J. Bend of trap to dishwasher Branch Tailpiece

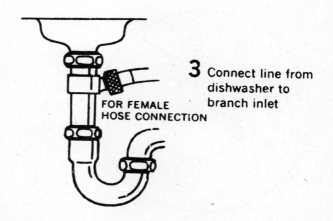

FOR FEMALE HOSE CONNECTION

3 Connect line from dishwasher to branch inlet

Troubleshooting Chart

PROBLEMS	CAUSES	REPAIRS
Faucet Drips	1. Faulty washer.	1. Replace the washer. For single-handled faucets, install all parts in repair kit.
	2. Improper seat.	2. Use reseating tool to grind seat even, or replace seat.
Hot water slows to trickle	1. Washer expands when hot.	1. Replace with proper nonexpanding washer.
Leaks around handle	1. Packing nut loose.	1. Tighten the packing nut.
	2. Packing not adequate.	2. Replace the packing.
Leaks around spout	1. Faulty O ring.	1. Replace the O ring.
Handle hard to turn; squeals when turned	1. Spindle threads binding against threads in faucet.	1. Lubricate spindle threads with petroleum jelly, or replace spindle.
Drains overflowing	1. Clogged pipes.	1. Use plumber's friend, chemicals, aerosol, or snake to remove blockage.

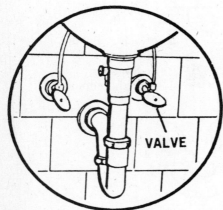

1 *To repair a leaky faucet, first turn off the water at the supply stop valve.*

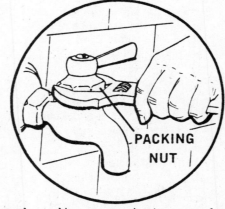

2 *Remove the packing nut, turning it counterclockwise.*

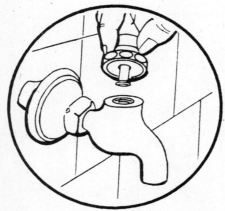

3 *Remove the stem or spindle from the body of the faucet.*

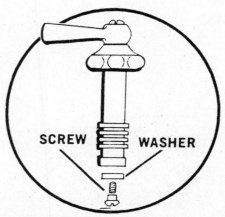

4 *Check the washer at the bottom of the stem; if it is worn or damaged, remove the screw holding it in place.*

Important Facts About Bathtub Installation

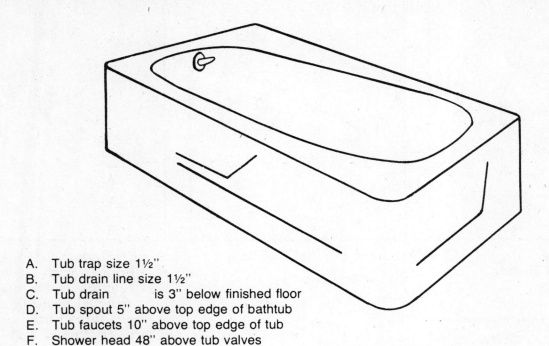

A. Tub trap size 1½"
B. Tub drain line size 1½"
C. Tub drain   is 3" below finished floor
D. Tub spout 5" above top edge of bathtub
E. Tub faucets 10" above top edge of tub
F. Shower head 48" above tub valves

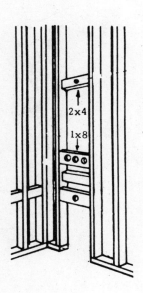

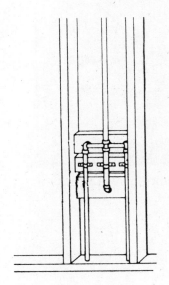

Front view of tub and shower valves.
Space holes in the mounting boards 4"
on each side of center hole.

Back view of tub and shower valves.
Use a 1" x 6" flush with studs to mount tub valves into.
Drill ¾" holes for shower head and tub valves.

REPAIRING TUB AND SHOWER VALVE

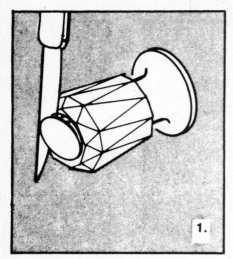

Turn off water.
Pry off Index Button.
Remove Screw

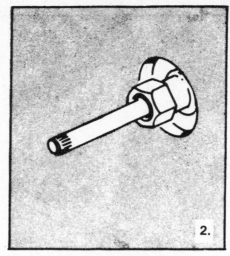

Lift off Handle   Unscrew
Flange and Nipple

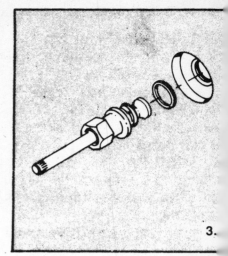

Prior to removal of the Stem
and Bonnet assembly for ho
and cold water, **be sure stem
is in full ON position and
water is turned off.** Unscrew
Stem and Bonnet assembly.

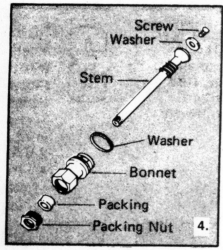

Unscrew Packing Nut from
Bonnet.

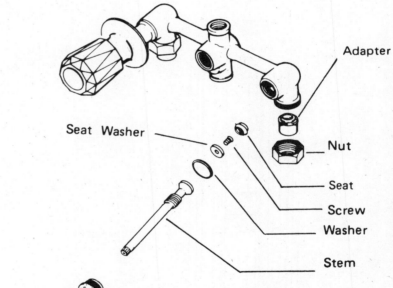

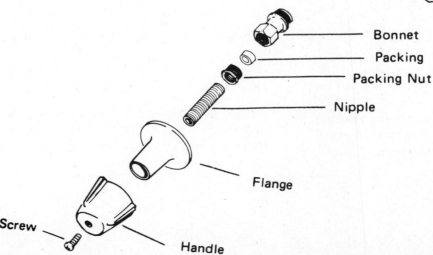

Plumbing Rough-in Measurements

Toilets

1. Toilet drain from center of drain to finished wall, 12".

2. Toilet water supply from finished floor 6" high.

3. Toilet water supply, 6" to the left of the center of toilet drain as you face drain.

4. Water supply: 2" out from finished wall.

Washbasins

1. Basin drain line from finished floor: 16" high.

2. Basin water supply from finished floor: 18" high.

3. Basin water supply: 4" to the left of drain and 4" to the right at 22" high.

4. Basin trap size: 1½". Use a 1½" x 1¼" reducing slip nut and washer to convert to 1½" drain line.

5. Washbasin line size: size 1½"

6. Water supply: 2" out from finished wall.

7. If wall-hung basin is used, put top edge of basin hangers 33" up from finished floor.

Kitchen Sinks

1. Kitchen sink drain with garbage disposal: 16" high from finished floor.

2. Kitchen sink without garbage disposal: 16" high from finished floor.

3. Sink trap size: 2".

4. Sink drain line size: 2".

Bathtubs

1. Bathtub trap size: 1½" P trap.

2. Tub drain line size: 1½".

3. Tub drain in floor: 3" below finished floor level.

4. Tub spout 5" above top edge of bathtub.
5. Tub faucets 10" above top edge of bathtub.

    (Check local plumbing codes).

Water Supply

Soldering
Solder Type Fittings
Pipe
Flare Type Fittings
Water Supplies

Information Chart on Copper Tubing

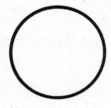

Type M Copper
Refer to local plumbing codes.
(Not permitted in BC)

Type L Copper

Type K Copper

Above is the approximate wall thickness of the various types of copper pipe.

Information on Copper	Type M Refer to local plumbing codes.	Type L	Type K
Where Used	In Homes Above Ground	In Homes Above and Below Ground Level	Below Ground Level
Pressure Rated Approximate	415 PSI	525 PSI	625 PSI
Color Coded	Red Stripe on Side	Blue Stripe on Side	Green Stripe on Side

Using a piece of string, wrap it around the tube and use chart to determine the size of copper you need.

1/2 " |————————————————|

3/4 " |———————————————————————|

Facts on Soldering

1. A simple explanation of soldering: heating the copper to a given temperature draws the solder in and around the fitting.

2. Use lead free solder (check with your local plumbing store on this).

3. Use a good non-acid flux.

4. If you overheat the solder joint, just reclean and solder the joint over again.

5. One pound of solder will do approximately 70 — ½" Fittings, or 40 — ¾ Fittings.

6. A two-ounce can of flux will do about 100½" fittings or 50¾" fittings.

7. When soldering overhead, the solder will draw into the joint the same as if you were soldering in front of you. Solder will flow upward.

8. Solder will not seal around the joint if there is even one drop of water in the line; if water is in the line, the solder will form little balls of solder instead of sealing or flowing.

9. If you are soldering a water line and there is a small amount of water still seeping into the line, preventing the solder joint from sealing — stuff some fresh bread (no crust) into the line; this will give you the few seconds you need to solder the line. When finished, just turn the water lines on and off a few times; this will dissolve the bread and flush it through the lines without any problems.

10. Use nylon or dialelectric unions when joining iron pipe to copper tubing.

11. There are three main things that will cause leaky solder joints. They are: (1) water — any amount — even one drop; (2) dirty pipe and fittings; (3) over-heating.

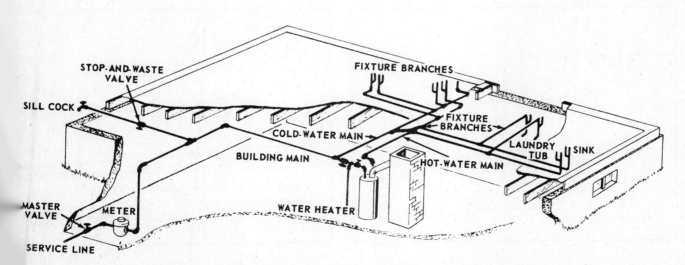

A TYPICAL WATER-SUPPLY SYSTEM FOR ONE BATHROOM AND A KITCHEN

The Six Simple Steps of Soldering

CLEAN THE OUTSIDE OF COPPER TUBE WITH FINE STEEL WOOL OR EMERY CLOTH TO A BRIGHT FINISH. **1**

CLEAN THE INSIDE OF COPPER FITTING WITH FINE STEEL WOOL OR EMERY CLOTH TO A BRIGHT FINISH. **2**

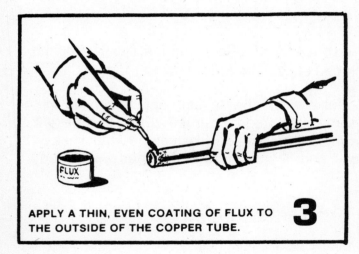

APPLY A THIN, EVEN COATING OF FLUX TO THE OUTSIDE OF THE COPPER TUBE. **3**

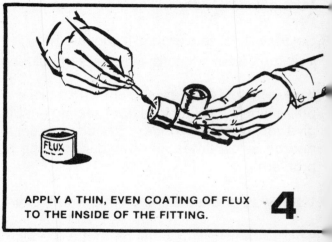

APPLY A THIN, EVEN COATING OF FLUX TO THE INSIDE OF THE FITTING. **4**

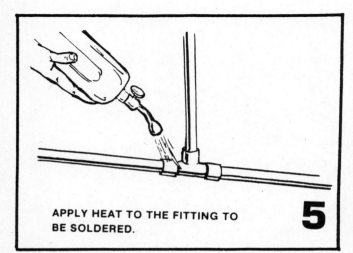

APPLY HEAT TO THE FITTING TO BE SOLDERED. **5**

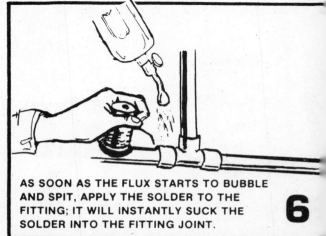

AS SOON AS THE FLUX STARTS TO BUBBLE AND SPIT, APPLY THE SOLDER TO THE FITTING; IT WILL INSTANTLY SUCK THE SOLDER INTO THE FITTING JOINT. **6**

Typical Water Supply for a Two-Story House

Solid line — cold water
Dotted Line — hot water

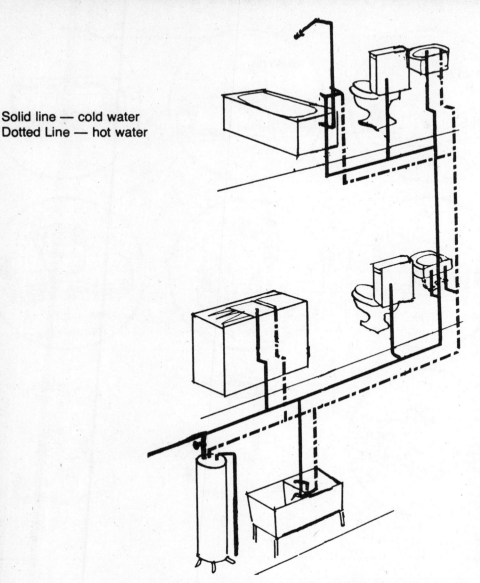

Material List for Two-Story House

¾" Material Needed	All Solder Type		½" Material Needed	All Solder Type
90 ft.	¾" rigid copper pipe		120 ft.	½" rigid copper pipe
1	¾" stop and waste valve, sweat type		45	½" 90° ells
1	¾" sweat compression stop		9	½" 45° ells
15	¾" 90° ells		9	½" male adaptors
10	¾" x ¾" x ½" reducing tees		6	½" female adaptors
5	3/4" x 1/2" x 1/2" reducing tees		9	½" couplings
3	¾" couplings		18	½" tees
3	¾" caps		9	½" caps
2	¾" dialetric unions		24	½" pipe hangers
3	¾" caps		1	propane torch kit
15	¾" pipe hangers		1	tube cutter
2 lbs.	lead free solid wire solder		1	acid brush
1	½-pint can, non-acid flux		1 lb.	lead free solid wire solder
1 pkg.	fine steel wool		1	¼-lb. can, non-acid flux
			1 pkg.	fine steel wool

Shock arrestors may be required for local plumbing codes.

Copper Sweat Fittings

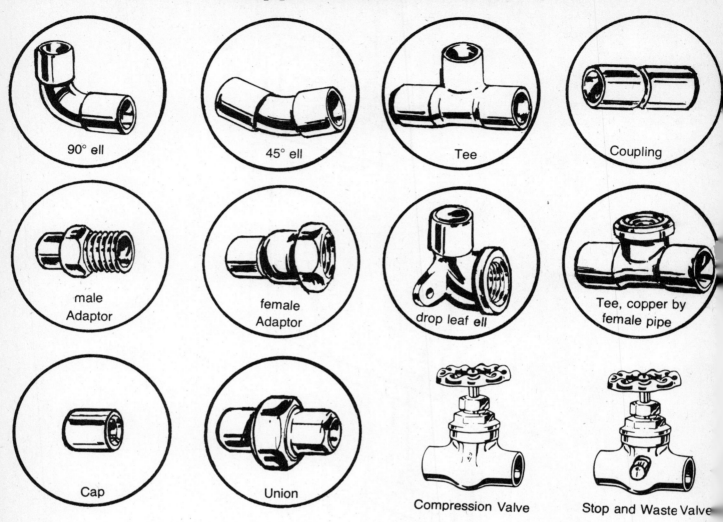

90° ell

45° ell

Tee

Coupling

male Adaptor

female Adaptor

drop leaf ell

Tee, copper by female pipe

Cap

Union

Compression Valve

Stop and Waste Valve

There are four main water line reducing tees. Reducing tees are figured on an A-B-C method. Examples below:

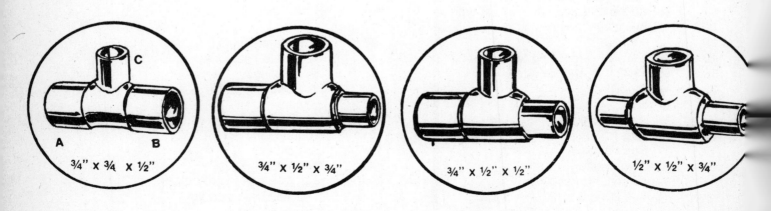

3/4" x 3/4" x 1/2"

3/4" x 1/2" x 3/4"

3/4" x 1/2" x 1/2"

1/2" x 1/2" x 3/4"

How to Flare Underground Copper Tubing

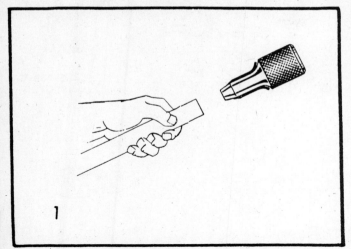

Flared Joints

Impact or screw-type tools are used for flaring tube. The procedure for impact flaring is as follows:

1. Cut the tube to the required length.

2. Remove all burrs. This is very important to assure metal-to-metal contact.

3. Slip the coupling nut over the end tube.

4. Insert flaring tool into the tube end.

5. Drive the flaring tool by hammer strokes, expanding the end of the tube to the desired flare. This requires a few moderately light strokes.

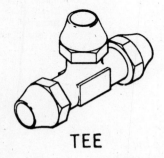

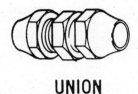

TEE

UNION

90° ELL

Male

Female

COPPER TO STEEL ADAPTER

FLARING TOOL

FLARE TYPE FITTINGS

FLARED TUBE FITTINGS

FLARE NUT

FLARE UNION

FLARE ELBOW

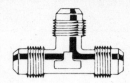

FLARE TEE

FLARE FEMALE CONNECTOR

To Install:

(1) Slide flare nut on tubing.

(2) Flare Tubing

(3) Screw flare nut on fitting—this seals the flared end of the tubing between the nut and the fitting.

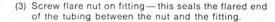

FEMALE ELBOW

FLARE MALE CONNECTOR

COMPRESSION TUBE FITTINGS
(Do NOT flare tubing)

COMPRESSION SLEEVE—

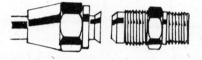

COMPRESSION NUT

COMPRESSION UNION— Both Ends Compression Tube

COMPRESSION MALE CONNECTOR— Compression Tube by Male Pipe Thread

COMPRESSION FEMALE CONNECTOR— Compression Tube by Female Pipe Thread

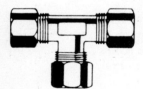

COMPRESSION TEE— Compression Tube All Ends

COMPRESSION ELBOW— Compression Tube by Compression Tube

COMPRESSION REDUCING SIZE UNION Both Ends Compression Tube

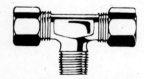

COMPRESSION BRANCH TEE— Compression Tube by Compression Tube by Male Pipe Thread

COMPRESSION MALE ELBOW— Compression Tube by Male Pipe Thread

To install:

(1)   Slip tubing through nut and sleeve.

(2)   Insert tubing into fitting up to shoulder.

(3)   Tighten nut hand-tight; then using a wrench, turn one-half turn, then turn on the water. If it drips, you turn the nut until it seals.

        Do not over-tighten.

Information On A.B.S. Plastic Pipe

1. A.B.S. is from the thermoplastic family; it stands for Acrylonitrite-Butadiene-Styrene.

2. Most manufacturers guarantee A.B.S. for 50 years; tests prove it will last almost indefinitely.

3. Use ABS primer or cleaner if required by glue manufacturer.

4. ABS will not corrode, rot, or lime-up. ABS glue melts the pipe to the fittings in seconds.

5. One pint of glue will do approximately 45 1½" joints, 35 2" joints, 20 3" joints, or 15 4" joints.

6. A.B.S. burns at 840°; wood burns at 320°; A.B.S. is definitely not a fire hazard.

7. On threaded A.B.S., use teflon tape.

8. The depth that pipe goes into fittings is as follows: 1½"=¾"; 2"=⅞"; 3"=1½".

9. Once the A.B.S. glue has set, you cannot take the joint apart; if a mistake is made, you have to cut out the mistake and put in a new fitting.

10. On A.B.S. pipe, use plumbers tape or pipe hangers at least 5 ft. apart.

11. The outside diameter of A.B.S. is the same as steel pipe, or cast iron drainage pipe; therefore, if you want to connect them, just use a no-hub coupling to put them together.

12. If you want to connect 1½" A.B.S. pipe and 1½" steel pipe, a no-hub coupling will do the job.

3. If you want to reduce 4" cast iron pipe to 3" A.B.S. pipe, you need a 4" x 3" no-hub reducer, one 4" no-hub coupling, and one 3" no-hub coupling. Band them together. Remember: the outside measurements on plastic and iron pipe are the same.

Assembling No-Hub Pipe

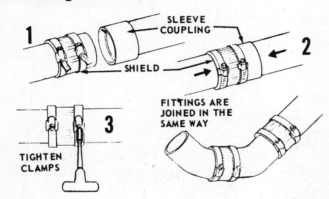

SLEEVE COUPLING

SHIELD

TIGHTEN CLAMPS

FITTINGS ARE JOINED IN THE SAME WAY

ABS Or PVC Drainage Installations

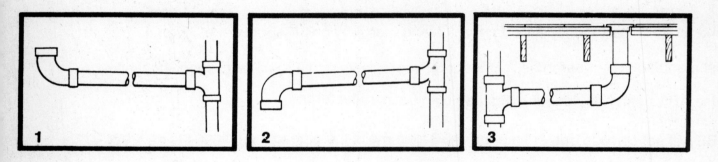

1. All fittings are of the drainage type which provides proper pitch for waste flow. Drain lines should be pitched ¼" per foot of length.

2. When drainage fittings are used on revent lines the fittings are inverted.

3. When installing closet flange refer to your toilet installation manual for rough in distance from finished wall to center line of waste outlet. Diameter of floor opening is five inches.

   The 4" x 3" reducing closet floor flange must rest on the surface of the finished floor.

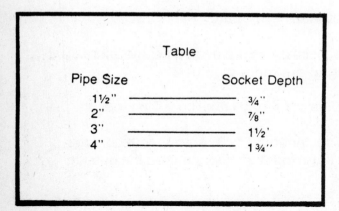

Table

Pipe Size	Socket Depth
1½"	¾"
2"	⅞"
3"	1½'
4"	1¾"

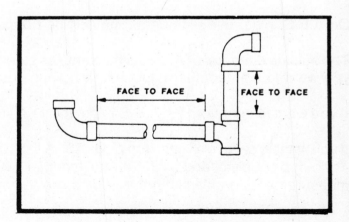

FACE TO FACE

FACE TO FACE

When cutting plastic pipe: (1) Measure the plastic fittings from face to face; (2) Allow the pipe to fit into the socket of the fittings to the depth indicated in the above chart.

VARIOUS CLEANOUT ARRANGEMENTS

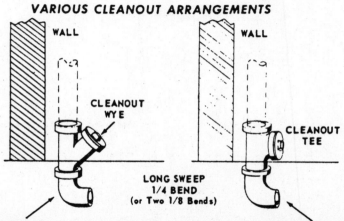

WALL

CLEANOUT WYE

WALL

CLEANOUT TEE

LONG SWEEP ¼ BEND (or Two 1/8 Bends)

See local plumbing codes. (Two 45° bends are required in BC)

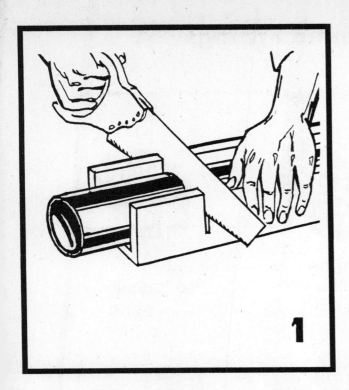

Step 1 Cut pipe with fine-tooth handsaw or hacksaw.

Step 2 With a piece of cloth, clean the inside of the fitting and the outside of the pipe.

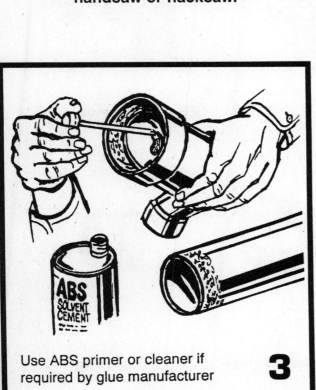

Use ABS primer or cleaner if required by glue manufacturer

Step 3 Apply a full, even coating of glue to the inside of the fitting and to the outside of pipe

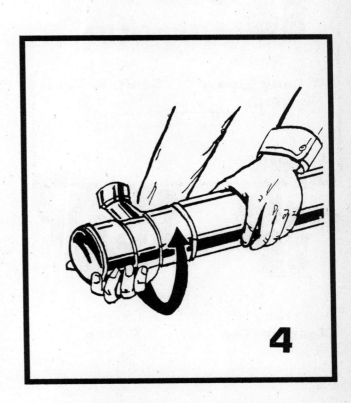

Step 4 Slip fitting over pipe, giving it a quarter of a turn; hold for a few seconds until glue sets.

A.B.S. or P.V.C. Plastic Fittings

Hub x Hub
22½° Elbow
1/16 BEND

Hub x Hub
45° Elbow
⅛ BEND

Hub x Hub
60° Elbow
1/6 BEND

Hub x Hub
90° Elbow
¼ BEND

Hub x Hub
90° Long Elbow
¼ BEND

All Hub
Sanitary Tee

Hub x Hub
45° Wye

All Hub
Wye & ⅛ Bend

Hub x Hub x FPT
Cleanout Tee

Hub x Hub
P-Trap

Hub
Closet Flange

Hub x Hub
Coupling

Hub   Female end of plastic pipe
Spigot   Male end of plastic pipe
F.P.T.   Female pipe thread
M.P.T.   Male pipe thread

A.B.S. or P.V.C. Plastic Fittings

All Hub
Sanitary Tee
(w/Right Side Inlet)

All Hub
Sanitary Tee
(w/Left Side Inlet)

All Hub
Double Sanitary Tee

SP x Hub
90° ¼ Bend
Street Elbow

SP x Hub
45° Street Elbow

Hub x Hub
Reducing
Sanitary Tee

MPT
Plug

Hub x Hub
Reducer

MPT x Hub
Male Adapter

SP x FPT
Spigot x Female
Pipe Adapter

FPT x Hub
Female Adapter

H	Hub, female end of pipe
S.P.	Spigot, male end of pipe
M.P.T.	Male pipe thread
F.P.T.	Female pipe thread
Adaptor	Adapts pipe to plastic

Facts on A.B.S. or P.V.C. Plastic Pipe and Fittings

There are two main types of ends on plastic pipe fittings — spigot end and hub end. Example below:

Spigot end

Hub end

Plastic Adaptors

An adaptor is pipe thread on one side of the fitting and plastic pipe on the other side. There are four different adaptors: male and female spigot, and male and female hub. They come in four sizes: 1½", 2", 3", 4". Examples below:

Spigot by male pipe

Spigot by female pipe

Hub by male pipe

Hub by female pipe

Reducing Plastic Fittings

Reducing tees are figured from end to end to middle. Example above is called a 3" x 3" x 2" reducing tee.

Some people refer to a reducing tee in a 1-2-3 method. Example above.

Reducing wyes are figured the same as reducing tees. The above is called a 2" x 2" x 1½" wye.

Plastic Vent and Drain System for
Tub — Basin — Toilet

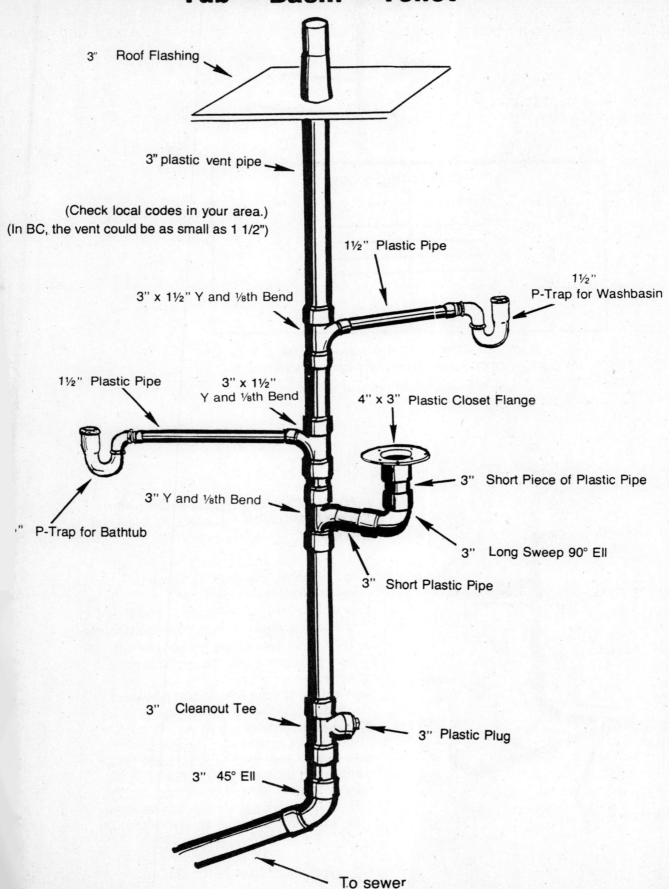

3″   Roof Flashing

3″ plastic vent pipe

(Check local codes in your area.)
(In BC, the vent could be as small as 1 1/2″)

1½″ Plastic Pipe

1½″
P-Trap for Washbasin

3″ x 1½″ Y and ⅛th Bend

1½″ Plastic Pipe

3″ x 1½″
Y and ⅛th Bend

4″ x 3″ Plastic Closet Flange

3″ Short Piece of Plastic Pipe

3″ Y and ⅛th Bend

″ P-Trap for Bathtub

3″ Long Sweep 90° Ell

3″ Short Plastic Pipe

3″   Cleanout Tee

3″ Plastic Plug

3″   45° Ell

To sewer

Plastic Pipe Vents and Cleanouts

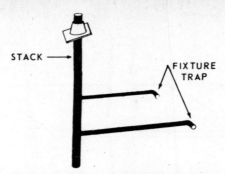

STACK

FIXTURE TRAP

BC plumbing codes.	
PIPE SIZE	**MAX. DRAIN LENGTHS**
1½"	5' ft.
2"	5' ft.
3"	6' ft.
4"	10' ft.

*MAX. HORIZONTAL DRAIN LENGTHS FOR
VARIOUS PIPE SIZES. WITHOUT REVENTING*
(Check local codes in your area.)

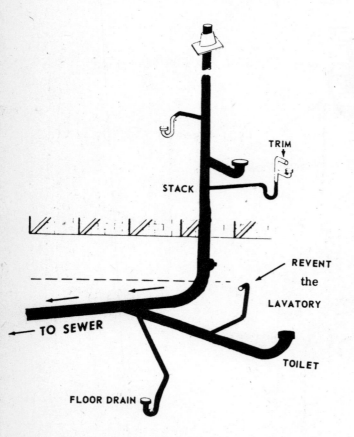

TRIM

STACK

REVENT the LAVATORY

TO SEWER

TOILET

FLOOR DRAIN

INFORMATION ON TOILETS, WASH BASINS, SHOWER AND FLOOR DRAINS IN THE BASEMENT

1. All drain lines should be pitched ¼" per foot of length, this equals 1" pitch to every 4 ft. of drain length.
2. Drain lines should be installed in wye fittings where you connect drain line together.
3. Toilet drain lines should not be over 10 ft. from main sewer line, if over 10 ft. it has to be vented.
4. Wash basin drain lines should be vented when installed in the basement.
5. Shower drain lines should be 2" in size, also need separate vent in basement.
6. Floor drain must be 3", a minimum of 18" in length and graded at 1/4" per foot but not exceeding the diameter of the pipe.
7. All vents must rise at least six inches above the flood level rim before the dry vent can be run on the horizontal.

(See local plumbing codes).

How a Plumbing System Works

Whether you are installing a whole house full of plumbing or simply adding some new fixtures, you must have a source of water and pipes to carry it to the fixtures. This water-supply system must be adequate to:

1. Assure you of pure water for drinking.

2. Supply a sufficient quantity of water at any outlet in the system — at correct operating pressure.

3. Furnish you with hot or cold water, as required.

In most incorporated areas the water source is a public or privately operated water "works" from which purified water is distributed through mains to which each user can be connected by arrangement with the proper authorities.

A SAFE DRAINAGE SYSTEM

Drainage (strictly controlled by code in most localities) is the complete and final disposal of the waste water — and of the sewage it contains. A drainage system, therefore, consists: 1) Of the pipes that carry sewage away from the fixtures, and 2) Of the place where the sewage is deposited. You may empty sewage into a city sewer, into a properly constructed septic tank or (in a few cases) into a cesspool.

To be safe, your drainage system has four simple requirements:

1. Pipes must be pitched (sloped) for gravity flow of the water downhill — all the way to final disposal.

2. Pipes must be fitted so that gases cannot leak out.

3. The system must contain vents to carry off the sewer gases to where they can do no harm.

4. Each fixture that has a drain should be provided with a suitable water trap — so that water standing in the trap will seal the drain pipe and prevent gases from backing up into the room.

5. Re-vents should be provided wherever there is danger of syphoning the water from a fixture trap or where specified by your code.

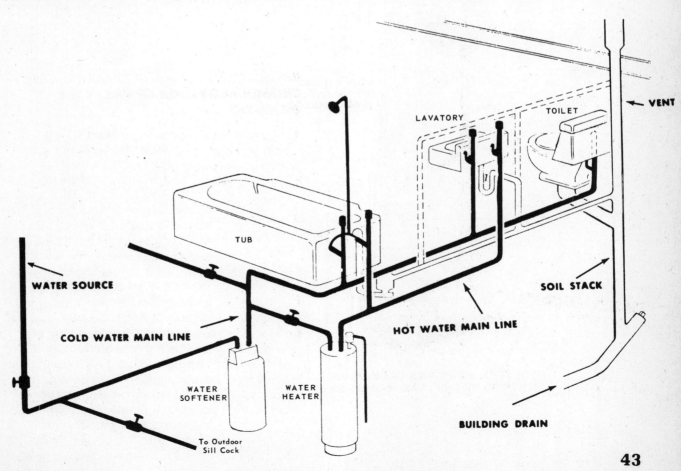

How A Trap And Vent Works

Traps.

A trap is a water-filled, U-shaped pipe that will allow water and wastes to pass through, but prevents gases and vermin inside the DWV system from slipping backwards into the house. Every appliance and every fixture must have a trap. Without traps, your house soon would smell like the inside of a sewer.

HOW A TRAP WORKS

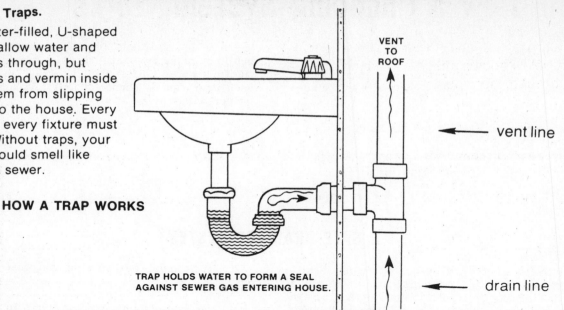

VENT TO ROOF

vent line

drain line

TRAP HOLDS WATER TO FORM A SEAL AGAINST SEWER GAS ENTERING HOUSE.

HOW A VENT WORKS

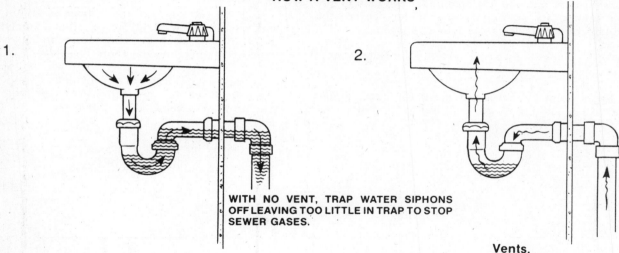

1.

WITH NO VENT, TRAP WATER SIPHONS OFF LEAVING TOO LITTLE IN TRAP TO STOP SEWER GASES.

2.

3.

WITH VENT, AIR RUSHES IN TO PREVENT SIPHONING OF TRAP, GAS SEAL REMAINS INTACT.

Vents.

water rushing along by gravity through a pipe creates a suction or vacuum at the high end of the pipe above it. This is called siphon action. Siphon action is powerful enough to suck all the water out of a trap and leave it nearly dry, as happens in a toilet bowl after a flush. But, since other fixtures and appliances aren't designed to replace siphoned-off trap water as a toilet is, some means of preventing trap siphoning must be built into the DWV system. This is accomplished by venting the system to outside air. Venting inside the house would work too, but then sewer gases would escape into the house. Thus, venting is done outside above the roof.

Pipe Sizing Chart

The chart below shows the size and type of drainage pipe used in today's plumbing. Place a piece of string around the pipe and use this chart to tell you the size you need or have now.

Size and Type of Pipe	Length of String
1 1/2 " Galvanized	6"
1 1/2" Plastic	6"
2" Galvanized	7 1/2"
2" Plastic	7 1/2"
2" Cast Iron	7 1/2"
3" Plastic	11"
3" Cast Iron	10 3/4"
4" Plastic	14 1/4"
4" Cast Iron	13 1/2"

To connect plastic pipe to either galvanized or cast iron pipe, use a no-hub coupling, which connects pipes together with no threads needed.

All that is necessary now is to slide the rubber coupling on both pieces of pipe. Slip the band over and tighten them together with a screwdriver.

No-hub couplings come in 4 popular sizes: 1½", 2", 3", and 4".

Galvanized Pipe Fittings

90° Elbow

45° Elbow

Street 90° Elbow

Tee

Reducing Tee

Union

Coupling

Reducer

Bushing

Cap

Plug

Nipple

P.V.C. Fittings and Pipe

90° ELBOW
Slip x Slip

90° ELBOW
Slip x Thread

45° ELBOW
Slip x Slip

TEE
Slip x Slip

TEE
Slip x Thread

MALE ADAPTER
Slip x Thread

SIDE OUTLET
ELBOW
Slip x THREAD

COUPLING
Slip x Slip

BUSHING
Slip x THREAD

CAP
Slip

PVC PIPE

CLASS 200 PVC PIPE

SCH. 40 PVC PIPE

GLOSSARY

Rough-in: All the water supply or drain lines that are behind the finished walls or below the finished floor.

Stack: The main vent in the bathroom, sized to local codes; also it is the vent pipe that sticks out through the roof of the house.

Traps: Plumbing fittings that are required under all plumbing fixtures; traps have a built-in water level in the bottom of the trap that prevents sewer gas from coming into the house. Toilets are the only plumbing fixtures that have their own built-in trap.

Vent: The pipe that carries away sewer gases and odors; also, the vents let air into the drain lines, letting the pipes drain without air locking.

Trim: Trap, supply lines, strainers, etc.

Spigot: The plain end of a pipe that goes into a hub fitting.

Adaptor: A fitting that connects two different sizes or types of pipe or fittings. In this book it refers to adapting plastic to cast iron.

DWV: Means Drainage Waste Vents. These fittings incorporate the recommended drainage pitch of ¼" to the foot.

A.B.S.:   From the thermoplastic family, it is an abbreviation of Acrylonitrite-Butadiene-Styrene.

P.V.C.:   From the vinyl chloride family — Polyvinyl chloride.

Toilet: Thomas Crapper invented the first practical flush toilet in England in 1870.

M.P.T.:   Male pipe thread.

F.P.T.:   Female pipe thread.

M.S.P.S.:   Male standard pipe size.

F.S.P.S.:   Female standard pipe size.

N.P.S.:   National pipe size.

Male: End of pipe which inserts in the female fitting.

Female: A pipe fitting that receives a male pipe.

Caulking: A method of sealing; also, caulking is referred to as a material, like oakum and lead.

GLOSSARY

Clean-out Plug: A plug that is used to clean out plugged drains; installed in tees and Y's.

Closet Screw: A brass-plated lag screw designed to expand metal anchors; also used to screw the toilet base to wood.

Closet Bolt: A special bolt designed to fit in a closet flange and hold toilet base to floor.

Wet vent: A wet vent is both a vent and drainage line from any fixture with a trap arm or fixture drain which does not exceed 2" in size.

O.D.: Outside diameter.

Stub: Pipe or tubing sticking out or up from wall or floor.

I.D.: Inside diameter.

Fixture: Basin, toilet, sink, etc.

Popular Plumbing Tools

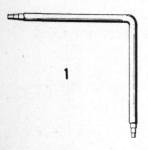

1

2

3

4

5

7

8

6

1. Seat wrench — for removing faucet seats.

2. Deep sockets — for removing tub and shower stems, etc.

3. Basin wrench — to reach basin and sink faucets in hard-to-get-at areas.

4. Closet auger — for toilets.

5. Flaring tool — to flare tubing.

6. Tubing cutter — to cut tubing.

7. Bending spring — to bend tubing.

8. Hand-type flaring tool — to flare large-size tubing.